The World of Robotics

Jordan Larsen

Copyright © 2012 Jordan Larsen

All rights reserved.

ISBN-10:147522785X
ISBN-13:978-1475227857

Note from the Author

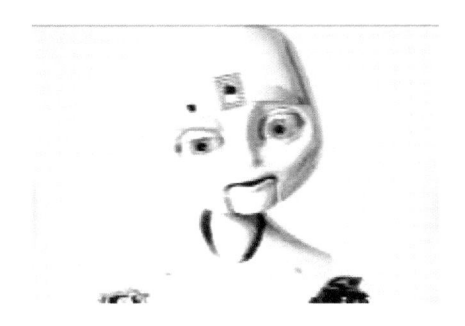

Originally, this project started in the Spring of 2003 as a research paper for my Physical Science Class in college. My professor, Richard Brill called it, "The Breakfast of Champions." Now I would like to take this opportunity to share it with those of you who dream of becoming the next great inventors and engineers of the future. What follows is an in-depth look at the history of robots and how the imagination of science fiction authors and filmmakers has given rise to some invaluable creations of technology that are being employed in the workplace as loyal servants and versatile lifesavers in some of the world's most dangerous professions. Most of you might have some pretty strong impressions of robots as *Transformers* and the *Terminator*, but the ones you'll meet in this book, are more real than anything Hollywood could ever envision on the big screen.

I have gone to great lengths to ensure that all historical facts highlighted in this book are accurate. If there are any inaccuracies about a particular robot or person described here I apologize for any unforeseen errors noted by the reader.

ACKNOWLEDGMENTS

This book would not have been possible without the help and invaluable support of some very important people. I would like to take this opportunity to thank Richard Brill, Professor of Physical Science at Honolulu Community College (HCC) and the University of Hawaii (UH) for giving me the freedom to write about this amazing topic. I will always be grateful for your kind words of praise and for calling my work "The Breakfast of Champions."   Thank you for making the topic of Physical Science fun given your wit and humorous anecdotes. A very special thank you goes out to my Dad, Colonel Glenn J. Larsen, United States Air Force (RET) for sending me helpful links on the web during my initial research phase of this project and helping me with the long editing process toward the end. Lastly, a very dear person and confidant in my life is my Mom, Annalissa Larsen, who took the time to read the finished copy of my research paper and gave me words of encouragement.

CONTENTS

	Acknowledgments	iv
	Prologue	1
1	Origins of Robotics	3
2	Robots in the Workplace	6
3	Robots of Three Mile Island	9
4	Bomb Squad & SWAT Robots	13
5	Firefighting Robot: Fire Spy	20
6	Search and Rescue Robots	24
7	Robots in Space	29
	Epilogue	33
	References	35

Prologue

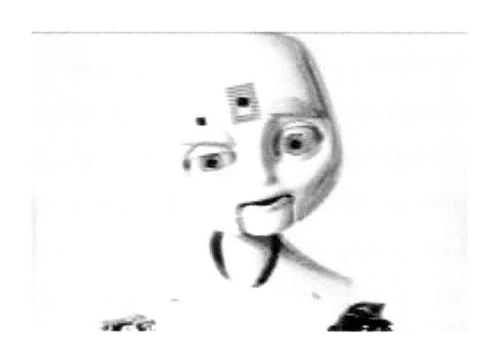

In 1934, a new word was added to the Webster's Dictionary. It was Robot, which means slave labor. Since that time, robots have more than lived up to that meaning. As technology becomes more advanced in the 21st Century, robots are now taking on new and bold endeavors in the workforce and even in the far reaches of our universe. Today, scientists and engineers are constantly finding new and more innovative ways to expand the capabilities and uses of robots. Since they first arrived on the scene in the 1960's, robots have been tasked with taking over more dirty, dull and dangerous jobs every year. They've been used in car factories, nuclear power plants to clean up and dispose of nuclear waste, as bomb diffusers in high-risk areas around the world to find and disarm deadly terrorist and criminal bombs and booby traps, as Industrial firefighters in putting out dangerous chemical plant and refinery fires in the United Kingdom and oil well fires in the desert of Kuwait, as search and rescue allies combing through the rubble at Ground Zero, as frontline combatants in the War on Terror in Afghanistan, as surgeons and nurses in hospitals and even as space explorers in outer space. These are the jobs that robots are beginning to get involved in, jobs that are either too dangerous or too time consuming for people to do.

1 ORIGINS OF ROBOTICS

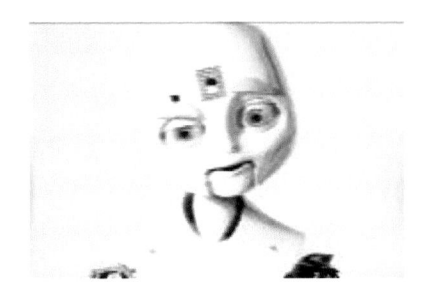

To understand how robots are being used in the workforce, one must first understand what they are and how they work. According to a 1997 Discovery Channel documentary, <u>Robots Rising</u>, "A robot is a machine that is capable of performing useful tasks over and over again and is capable of making decisions about its environment without constant human intervention. In other words, robots are the brainchildren of computers and power tools. There are three basic ways to make robots work. A Tele-operated robot has some computerized control at the robot end but a human operator makes virtually all of the decisions. With a Supervisory robot, the human operator makes key decisions about what the robot should do and the robot has enough intelligence to perform these tasks without constant guidance. An Autonomous robot is virtually freed of all human control." As Science Fiction films have preached for many years, this might not always be a good thing for humanity.

    The public's fascination with robots first got its major start with the advent of Science Fiction films in the 1920's and into the late 1980's with box office films such as, *Star Wars*, *The Terminator* and *Robocop*. Hollywood has often fantasized about man and machine relationships as well as the dark side of future robots like autonomous killing robots in the *Terminator*, which are devoid of remorse, loaded with firepower, hell bent on deceiving their creators and taking over the world. According to a recent documentary on the History Channel, <u>Robots</u>, this haunting obsession of robots having evil minds of their own and threatening to take over the world, first started with Frankenstein writer, Mary Shelly. In 1920, she envisioned a human-like machine that becomes alive and turns into a terrifying monster. Her frightening vision would become a classic novel and film that would become the standard for Science Fiction films for many years to come. However, a few friendly Science Fiction icons have found their way into our imaginations. R2D2 and C3PO, the intelligent and humorous robots from the 1970's film *Star Wars* 'lived' to serve their human counterparts. Ironically, the fantasy vision of human workers and sophisticated robots working together side by side was made a reality long before *Star Wars* ever hit the movie theaters.

2 ROBOTS IN THE WORKPLACE

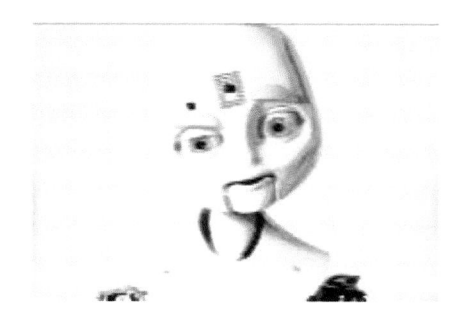

In 1961, machine manufactures George C. Devol and Joe Engelberger invented the first industrial robot to work alongside workers in the factory. The mammoth sized robot was named Unimate. It was intended to perform strenuous and often hazardous tasks in the factory of General Motors in Detroit. The Unimate is a supervisory robot. It would obey step-by-step instructions, which would be stored on a magnetic drum. With its four thousand pound arm and claw, it would stack and sequence hot pieces of die cast metal that it would remove from a hot furnace. While performing these tasks, an operator would be in a room across from where the robot was working. He would be sitting at a computer terminal control console with a joystick making decisions about what the Unimate would do in the factory. The price tag for the Unimate was an amazing $12,900. However, for those who envisioned and made it into a reality, such as, George C. Devol, it was more than well worth it. Interestingly enough, the Unimate marked another impressive milestone for its time. It was the first robot to ever appear on national television as a guest on the Tonight Show. The perception was, "If you could make it on the Tonight Show, you could make it anywhere." The Unimate would perform humorous antics such as, playing golf and even advertise Beer by picking up a bottle and pouring a glass of it on stage in front of a live audience. For its creator George C. Devol, the success of Unimate was more than he could ever have wished for and later on would form his own company called Unimation, which would produce future Unimates to work in the factory. However, it would take 14 years before his company would make a profit with its industrial robots. With robots such as the Unimate becoming instant hits in factories by the late 70's and early 80's, the idea of robots working in car factories to build and customize hundreds of cars on the assembly line everyday was already being well practiced in the United States. Due to the success of Unimate in the 1960's, by 1982 General Motors became the largest user of robots in the world. The franchise even caught on with the Japanese; who by 1990 had more than

40 companies producing robots, according to an online website, ibotz.com.

3 ROBOTS OF THREE MILE ISLAND

Photo courtesy of TEPCO.

The introduction of robots in the factory was a great relief for their human co-workers but imagine if you had to work in a nuclear power plant that is about to undergo a disastrous meltdown. This was the frightening situation that was faced by plant workers and managers at the infamous Three Mile Island nuclear power plant in Pennsylvania on the morning of March 28, 1979. According to a news article on the super 70's.Com website, "the Three Mile Island (TMI) Nuclear Generating station is located on 814 acres on an island in the Susquehanna River some 10 miles southeast of Harrisburg, Pennsylvania near some farmland. The incident occurred at four o'clock in the morning. The cooling system in the # 2 generator had failed during routine maintenance resulting in the reactor core being partially exposed, which led to some radioactive gases escaping into the containment section of the reactor building. Though some of this radiation was released into the surrounding area, no immediate deaths or injuries occurred. In the end, officials were able to restore enough coolant to the reactor core to prevent a complete meltdown and the #2 reactor at TMI was shut down permanently. The #1 reactor was also shut down and didn't resume operation until 1985."

In 1985, it was time for the cleanup operation to begin at TMI, but doing so presented a problem. Most of the area near the #2 reactor was contaminated and plant managers needed to assess what kind of damage had been done and find ways to safely repair it. This incident set the stage for a new breed of industrial robots, the hazardous duty robots. These are robots that go into places that are far too dangerous for people to work in. One of the first of three robots scheduled to venture into the forbidden environment of the Three Mile Island site was a small robot called the Remote Reconnaissance Vehicle (RRV). Supported by Carnegie Mellon University's Professor Red Whittaker and graduate student Jim Osborn, the RRV was lowered by crane into the contaminated reactor area. Once inside, the robot would use the crane cable as an emergency escape route in case anything went wrong exploring the damaged reactor area.

   According to a recent documentary on the History Channel, <u>Robots</u>, the RRV came equipped with a series of cameras to give plant managers and clean up personnel a good idea of the type of damage that had been done. The robot skillfully scanned the reactor of unit # 2 and also went down into the basement of the reactor area. The RRV is a Tele-Operated robot and the first of its kind to go into such a dangerous environment. As a result of the information that it relayed to Whittaker and others involved, two more robots would go into the TMI danger zone. The second robot would measure radioactive levels and another would decontaminate and dismantle the deadly radioactive site. According to an online news article at <u>post-gazette.com</u>, before the third robot, Workhorse, could go in and do its much needed work, plant officials decided to seal the damaged reactor before the robot even had a chance to go in. The Workhorse robot cost $2 million to build and TMI officials sold it back to Whittaker for one dollar. However, robots like the RRV and Workhorse helped pave the way for other hazardous duty robots such as, Rosie and Houdini. These two robots have different abilities but share the same goal which is to tackle the equally dangerous tasks of cleaning up nuclear waste as well as decontaminating and dismantling nuclear power plants. Rosie's job is to dismantle nuclear plants and Houdini's is to clean it up. According to a 1997 Discovery Channel documentary, Robots Rising, "Houdini works at the Carnegie Nuclear Weapons plant. When he lowers himself into a reactor area, his job is to clean up and store waste sludge that is so lethal it could kill most of the life on earth."

Working at a Nuclear Power plant might sound dangerous enough; but there's another occupation where robots have already well established themselves; particularly if your job description entails deactivating and removing bombs and deadly booby trap devices planted by terrorist and criminal bombers.

4 BOMB SQUAD & SWAT ROBOTS

PHOTO BY GLENN J. LARSEN AT US ARMY STRONG EXHIBIT, SAN ANTONIO, TEXAS, 2009.

A brief case left unattended in a crowded mall or airport terminal, an unmarked car parked at the end of a busy intersection, or a pipe bomb left inside a school locker – these are some of the all too familiar scenarios faced by the brave men and women of police and military bomb disposal squads. It's a business that requires a level head, the fine-tuned hand of a surgeon, decisive decision-making and an emotional patience beyond the ordinary. However, the job isn't fraught without extreme risk or personal tragedy. On New Year's Eve, 1982, two NYPD bomb squad technicians suffered critical and debilitating injuries while attempting to defuse a bomb. As a result, both officers suffer hearing loss and blindness due to their encounter with what experts call, "the coward's weapon of choice." In 2002, in a village in Afghanistan, four Army Corps of Engineers bomb disposal technicians were killed while trying to defuse a massive bomb. These are just some sobering examples of how potentially dangerous and life threatening the job of the bomb squad can be. According to a 1997 documentary, <u>The Future of Crime Fighting</u>, "In 1994 there were 2500 bombings and attempted bombings in the United States." Since that time, the number of calls that bomb squad members have received has more than doubled. According to a 1998 documentary, <u>On The Inside: Bomb Squad</u>, "more than 5,000 bombs in the U.S. and 450 bombs in Canada are defused or detonated every year." In many of these cases, bomb squad members often called upon an expendable mechanical ally, the bomb squad robot. With the constantly evolving skillful ingenuity of bombers and the ever-growing frequency of bombing attacks, the bomb squad robot is often the first choice and only savior for bomb squad technicians in high-risk areas around the world. One of the most widely used robots in the bomb disposal arsenal throughout the United States is the Andros Mark 5-A1 robot.

This 670-pound robot comes with a variety of tools and attachments for assessing and defusing potentially lethal devices. The Andros Mark 5-A1 is the largest and strongest robot in the bomb disposal business. According to an online site, Remotec.com, "Its 3.5mph speed, coupled with its unique articulated tracks allows the Andros Mark 5-A1 to rapidly maneuver over rough terrain and obstacles, climb stairs and cross ditches as wide as 24 inches. The vehicle is environmentally sealed to operate in any weather condition and in areas of high temperature and humidity." There is no cutting the red wire for this robot, the Andros Mark 5-A1 robot comes with a powerful dexterous claw for handling suspect packages or devices such as a pipe bomb, or an explosive device left inside a gym bag or briefcase. Its attachments also include an X-Ray Mount assembly to scan inside suspect packages, two cameras, one video and one Infrared, which allows the operator to navigate the robot from a safe distance. For tackling potentially dangerous packages or devices, the Andros robot comes with a powerful water canon known as a Disrupter, with a sophisticated laser targeting system. The disrupter performs very much like a gun. The long tube of the disrupter serves as the barrel; however, its payload is nothing more but simple tap water. Water is non-compressible and when it's fired out of the disrupter at high pressure it has the ability to penetrate any object or device. When the disrupter is mounted on the robot, it gives bomb squad members a distinct advantage. The robot can go down range into the danger zone and perform what bomb squad technicians' call, "Rendering it Safe." The render safe procedure essentially means either blowing up the device with a small amount of explosive or ripping it apart with the aid of a disrupter. The goal here is to safely disrupt or disconnect the fuse or the timer and render the device harmless. With the aid of its laser targeting system, the operator can fire the disrupter at the heart of a bomb. When the disrupter is fired, the package or device is ripped apart to pieces. The force of the water is so powerful that even motion sensitive devices are not safe from the effectiveness of the disrupter. According to a 1997 documentary, <u>On The Inside: Bomb Squad</u>, "The water strikes faster than the speed of motion sensitive triggering devices."

    For the human operators fortunate enough to have the opportunity to control the Andros robot, it's well worth the time and effort. The bomb squad can choose between two ways of controlling the robot. They can use either a compact Briefcase Control system or a large Stand-up Controller with several joy sticks and switches to operate many of the robot's different functions. Just like the RRV and the Workhorse robots, the Andros Mark 5-A1 robot is Tele-operated, which means that its actions can be controlled from a safe distance by the bomb squad. The squad has two ways of controlling the robot. They can choose between the uses of a Fiber Optic Cable Reel Assembly, which can allow the robot to operate as far away as 1000 feet; or they can use a Radio Control Assembly, which allows them to control the robot and also use its two-way microphone if necessary. However, because most criminal and terrorist bombs are radio controlled, most bomb squads prefer to use the Fiber Optic Control Cable to avoid setting off a device accidentally.

    The members of the bomb squad spend hundreds of hours training and perfecting their skills with the robot. For them, it's deadly serious business; operating the robotic arm, disrupter and cameras, takes great skill and precision. Having the right tools and the right training makes all the difference because in this job, the difference between success and failure is measured by one short fuse. Bomb squad robots like Andros and others like it have been credited with saving hundreds of lives. Today's bomb disposal robots cost anywhere from 125-300 thousand dollars. However, when it comes to defusing bombs, no one worries about the cost, robots can be replaced, human lives cannot. Many of today's police robots that were once used for bomb disposal are now finding their way into the realm of the SWAT team where they are being outfitted with various attachments such as shotguns, non-lethal beanbag weapons, high intensity lights, tear gas dispensers and sophisticated sensors that can detect various chemical agents such as VX gas or Sarin. It's as close to Robocop as you can get. While most hazardous duty robots are primarily used for bomb disposal, cleaning up nuclear waste and high-risk police situations, there's another occupation where robot technology is moving into; the world of the Industrial firefighter.

The British Army's Bomb Disposal Robot. The "Wheelbarrow" has seen action in Northern Ireland & is credited with saving hundreds of lives.

The World of Robotics

Foster-Miller's "Talon" Robot used for bomb disposal & HAZMAT.

Photo by Glenn Larsen at US Army Strong Exhibit, San Antonio, Texas, 2009.

I-Robot's Packbot at US Army Strong Exhibit, San Antonio, Texas, 2009.

Photo by Glenn Larsen.

5 FIREFIGHTING ROBOT: FIRE SPY

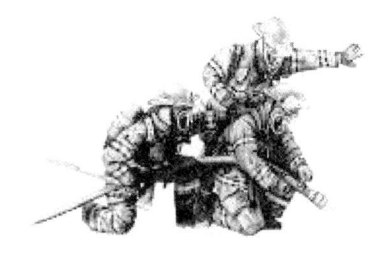

Imagine a huge chemical plant is on fire; you have toxic gases and hundreds of volatile chemicals to confront, one of which if caught by the fire, will release a deadly cloud of cyanide gas and thousands of innocent people in a nearby town are in harm's way. This was the nightmare faced by firefighters in Bradford, England on July 21, 1992. On that day, Bradford, England became the focus of one of the worst hazardous material disasters in British history. The plant not only housed several explosive and flammable chemicals but also held a 600-ton tank of Acrylo-Nitryl, a chemical so explosive that if ignited by the fire it would cause a horrific explosion that would take out half the town of Bradford and release a cloud of deadly cyanide gas. In one hour, 32 fire trucks and more than one hundred firefighters and hazardous material specialists are fighting the fire. Because of the volatile nature of the Acrylo-Nitryl, firefighters shower thousands of gallons of water to protect it from the heat of the fire. As a result of all the water being poured onto the fire, firefighters now have a new problem; the danger of toxic run-off water that eats away at their protective clothing. Fortunately, the run-off water isn't toxic but it's flammable. After three hours and thousands of gallons of water and foam, the blaze at Bradford chemicals is put out. Although, some areas of the plant had to be abandoned to the fire, the explosive Acrylo-Nitryl never got a chance to ignite and destroy the plant along with thousands of people. As a result of being exposed to deadly chemical fumes and toxic run-off water, 20 firefighters are injured and six hospitalized for damage to their lungs. Amazingly, there were no deaths as a result of combating the massive inferno at the Bradford chemical plant.

In response to the disaster at the Bradford chemical plant, the fire service of West Yorkshire, England pioneered the use of a revolutionary machine in June of 1999. Enter Fire Spy, the industrial firefighting robot.

Photo Courtesy of BBC Online

Its developers are the West Yorkshire fire service and the UK's JCB. The robot resembles the appearance of a Forklift except this robot has a powerful grabbing arm and claw for removing flammable and dangerous chemicals. Fire Spy has been built to withstand temperatures as high as two thousand degrees and the wiring has been upgraded to survive extreme temperatures. According to a BBC online article, Fire Spy costs $50,000 and like the Andros 5 robot, Fire Spy is Tele-operated. The Tele-operator can maneuver Fire Spy's claw by a series of joysticks and can see what he's doing by the use of a virtual reality headset that beams back video images from two cameras mounted on Fire Spy; one video and one Infra-red. Fire Spy can be controlled as far away as 100 meters or more than 300 feet. It even has a special hose attachment to spray water directly into the heart of a fire. However, like most industrial robots, Fire Spy is an experimental prototype; and after four years of service and 14 chemical fire incidents, it was retired from service. The reason was because it couldn't operate over rough terrain. If the robot were on a flat surface then it would work well but if there were cables and machine parts scattered around the area of the fire, then it would have problems operating. According to the West Yorkshire Fire Brigade Union, "There is no replacement for the firefighters themselves." If an expensive piece of equipment like Fire Spy doesn't see action for a long time or has problems operating in difficult terrain, then it's unreasonable for it to stay in service; and most departments can't afford to keep it. On the other hand, the introduction of Fire Spy is an excellent example of how robot technology, can safeguard many industrial firefighters lives. While industrial firefighting robots like Fire Spy help firefighters tackle industrial fires, there's another dangerous occupation where the success and use of robots has become well founded. This is the world of the Search and Rescue robot.

6 SEARCH AND RESCUE ROBOTS

PACKBOT ROBOT RUNS OBSTACLE COURSE AT ARMY STRONG EXHIBIT. SAN ANTONIO, TEXAS, 2009. PHOTO BY GLENN LARSEN.

The place is New York City, the date, September 11th, 2001. The peaceful routine of a quiet morning in lower Manhattan is shattered forever by the unthinkable. Two Boeing 767-passenger jet planes hijacked by terrorists crash into the World Trade Center towers. Hours later, both towers have completely collapsed; and what remains of the two 107-story office towers is a massive pile of twisted wreckage, plaster and pipes. As swarms of firefighters and other emergency personnel comb the wreck site for any survivors including their own fallen comrades, a different kind of rescuer arrives on the scene. The search and rescue robots; these mechanical allies came from both military and research institutes. One of the first agencies to provide robotic assistance was CRASAR or (the Center for Robot-Assisted Search and Rescue). CRASAR arrived at Ground Zero in just six hours of being alerted of the attack. Their robots went to work combing the wreckage, traveling into crevasses and other tight areas that were too dangerous or too difficult for firefighters and other rescuers to enter. Other teams from the military put their robotic resources to work within minutes of receiving their orders. Ground zero was the proving ground for many unique robots that were typically used in other dangerous professions such as bomb disposal, nuclear power plants and mine clearance work. According to the CRASAR rescue robots website, many of the robots utilized by rescue workers at Ground Zero ranged in size from that of a shoebox to a small brief case. Equipped with searchlights, sensors, video cameras and tank treads for maneuvering over piles of debris and cement rocks, the search and rescue robots methodically combed the wreckage for any signs of life.

    Just like Fire Spy and the Andros 5 robot, the search and rescue robots at Ground Zero are Tele-operated. And because of their size, operators can maneuver them through small gaps and openings very easily. According to Charles Werner, Firehouse.Com Techzone Editor, "These robot units are very versatile as they can manipulate their size and get into areas that are very confined and otherwise inaccessible by humans. They can be equipped with video cameras with infrared capability with the images sent to rescuers working on the surface via a cable or wirelessly. In a safe way, rescuers can determine the dangers and the necessity of further exploration." When all was said and done at the disaster site, over a dozen remote controlled robots were deployed at Ground Zero. Even though they were able to negotiate the mammoth pile and beam back haunting images of what was once the World Trade Center, no victims were found alive in the rubble. However, many more rescuers would have likely been injured or killed in the weeks following the attack, if not for the assistance of the military and research institute search and rescue robots. Robots like those used by CRASAR and the military at Ground Zero have also seen action in places like Turkey and India to provide support to rescuers in the aftermath of earthquakes. Not long from now, small versatile robots will be used to assist emergency first responders in situations like hazardous chemical spills, nuclear bomb incidents and even chemical or biological terrorist attacks.

With the success of small compact robots at Ground Zero, robot technology is beginning to expand into the realm of search and rescue. Search and rescue robots aren't just helping rescuers on the ground but they are taking to the air. The same robotic technology that is coming to the aid of search and rescue specialists all over the world is now being developed for high-rise rescue vehicles. In the future, sophisticated flying machines such as Israel's Urban Aeronautics X-Hawk will come to the aid of firefighters and civilians at high-rise fire incidents as both evacuation vehicles and medevac rescue vehicles. These vehicles have an original futuristic design. A cockpit for the pilot, a large platform for passengers and according to a Popular Science article, " The X-Hawk machine comes with ducted fans, which means that the machine can safely approach tight areas where helicopters can't go. For example, up against a high-rise building in New York City." You can see the X-Hawk concept in action at www.urbanaero.com.

In the near future, as robot technology becomes more highly advanced, search and rescue robots will be getting smaller and more mobile. According to a recent documentary, National Geographic Explorer: Secret Weapons, "Autonomous search and rescue robots the size of a dragonfly will be staged throughout major cities ready to response to a disaster; wired to a central command center, if an explosion occurs and a building collapses, dragonfly robots will sense the vibrations and be deployed to that area." The ability to sense vibrations is critical for search and rescue robots. If there's a secondary collapse, they can take evasive actions to avoid being crushed. This futuristic technology is invaluable for firefighters and other emergency specialists. By enlisting the aid of high-tech search and rescue robots, they will be minimizing the risk of human injury or death, so that others may live.

7 ROBOTS IN SPACE

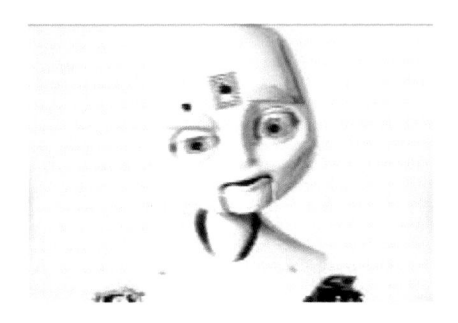

While today's workforce robots are being used in hostile or dangerous environments for safe guarding human lives, there's another more forbidding environment that has only been occasionally visited by humans and is not yet totally understood. Enter the realm of space travel and the world of the astronaut. The endless regions of space have often been called the final frontier and it's also the perfect setting for the next generation of robotic technology.

Meet DART (Dexterous Anthropomorphic Robotic Testbed).

Photo Courtesy of NASA JSC

He is one of the first robots developed by NASA to have the ability to assist astronauts in space. Here, DART is an entirely different robot in itself; apart from most tele-operated robots that work by use of joysticks and control panels, DART has the ability to be controlled by the body's natural movements. As a result, it can perform like no other robot. Its control category is classified as a Telepresence system, which means that it requires the presence of a human operator to mimic precise movements in order for the robot to perform various tasks. In order to operate the complex robot, the operator must use virtual reality gloves, sensors for his arms and a sophisticated headset. When the operator is ready, he essentially becomes the DART robot.

However, like most tele-operated robots, depth perception and judging distance is very difficult. When the operator slips on the virtual reality headset, he becomes totally immersed in the robot's world. After long periods of time, it can get very disorienting. People have become physically ill while being lost in DART's world. Aside from the dizzying aspect of it, the operator can't feel what the robot touches. However, the only way to overcome these obstacles is to constantly train and hone various skills. DART has already made wonderful progress in performing many useful tasks that would be required on a space shuttle or space station such as "tying knots and working with tether hooks that astronauts would use for space walks." Back here on earth, DART could prove to be very useful in some of the more delicate and dangerous jobs like construction, working in hospitals and bomb disposal work.

A Final Note from the Author

Since this writing in 2003, the Space Shuttle has sadly been retired from active service and the International Space Station (ISS) now has a robotic astronaut on board. Known as Robonaut 2 or R2 for short, it is designed to help ISS crews with complex tasks. You can find out more about this revolutionary robot at

http://robonaut.jsc.nasa.gov/default.asp.

NASA's "Robonaut" designed to help ISS crews with complex and dangerous tasks. Photo by Sarah Worthy.

EPILOQUE

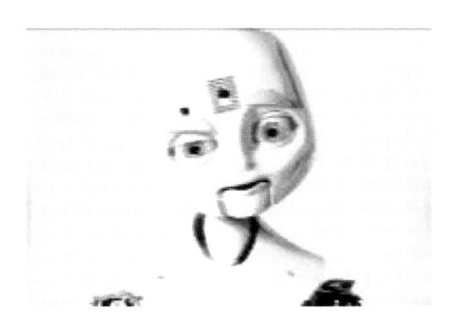

As technology continues to move further into the 21st Century, the use of robots has become so vast, and they're capabilities so evolved, that it's difficult to describe exactly every detail of their world. As my Physical Science professor Richard Brill once said, "Physical Science is like following a river." The science and technology of robotics, is very similar. Scientists and engineers find many different turns and encounter obstacles along the way while inventing robots, but out of those obstacles and winding turns, a monumental concept can emerge to better serve mankind. As we move further into the future, robots will continue to function as our loyal servants, our noble guides, and our valiant guardians while we continue to explore the far reaches of our own world and that of our distant universe.

REFERENCES

Advance Research & Robotics (Editors).. *The Start of a Revolution.* Retrieved April 5, 2003, from http://ar2.com/ar2pages/uni1961.htm

Army Technology (Ed.), *REMOTEC INC - EOD, Surveillance, Security and Hazardous Material Robots.* Retrieved from Army-Technology.com: http://army-technology.com/contractors/mines/remotec/

British Broadcasting Corporation (Ed.), *Robot Fire Fighter Unveiled.* Retrieved April 23, 2003, from BBC Online Network: http://www.bbc.co.uk/1/hi/sci/tech/281197.stm

Briggs, I. (2002, July 17). *Firefighting robot 'can go to blazes'.* Retrieved April 23, 2003, from http://www.thisisbradford.co.uk.bradford_district/archive/2002/07/17/brad_news08.int.html

Discovery Channel (1997, Spring). *Robots Rising* [Television series]. (Available from Discovery Channel, www.discoverychannel.com)

Discovery Channel (1998). *On the Inside: Bomb Squad* [Television series episode]. Montreal, Canada; Washington D.C.: Author.

History Channel (2003, January). *Robots* [Television series episode].Roger Mudd.

iBotz Online.. iBotz Online (Ed.), *A brief History of Robots.* Retrieved March 27, 2003, from iBotz Online: http://www.ibotz.com/html/CustSuptHistory.html

Kantra Kirschner, S., & Everett, J. (Eds.). (2003, February). How about an Urban Aerial Utility Vehicle. *Popular Science,* 13.

Kobell, R. (2000, February 13). Robots: Will they ever deliver on their promise? *Post-Gazette.* Retrieved from http://post-gazette.com/businessnews/20000213robots1.asp

Learning Channel (2001, Summer). *Into the Flames: the Poison Fire* [Television series episode]. Bradford, England; Kuwait; Texas City and Houston, TX: Author.

Learning Channel (1998, Summer). *The Future of Crimefighting* [Television series episode]. (Available from Richard Belzer)

LoPresti, R., & Gordon, L.. *Revolutionary Design of Special Forces and Hi-Rise Rescue Vehicles.* Retrieved May 11, 2003, from http://www.logovtol.com/

MSNBC (Producer). (2003, Spring). Secret Weapons. In *National Geographic Explorer.* Boyd Matsen.

Murphy, R.. R. Murphy (Ed.) & R. Murphy (), *Center for Robot-Assisted Search and Rescue CRASAR(tm)* [Data file]. Florida: University of South Florida. Available from CRASAR: http://www.csee.usf.edu/robotics/crasar/

NASA (Ed.), *Dexterous Anthropomorphic Robotic Testbed (DART).* Retrieved May 9, 2003, from DART: http://vesuvius.jsc.nasa.gov/er_er/html/dart/

NASA, JPL. (Ed.).. *URBIE Urban Robot.* Retrieved April 28, 2003, from Jet Propulsion Lab - California Institute of Technology: http://robotics.jpl.nasa.gov/tasks/tmr/homepage.html

NRC (Editors), Fact Sheet on the Accident at Three Mile Island. Message posted to U. S. Nuclear Regulatory Commission: http://www.nrc.gov/reading-rm/doc-collections/fact-sheets/3mile-isle.html

PBS (Authors), (1997, Fall). *Nova Bomb Squad* [Television series episode]. Northern Ireland:

Pennsylvania Firefighter Main Page (Ed.), *Extreme Firefighting.* Retrieved April 23, 2003, from Extreme Fire Equipment Home Page: http://www.pafirefighter.net/Extreme/Rainbow5/htm

Pope, J. (2002, July 30). Pvt. Robot Reporting, Sir! *Honolulu Advertiser.*

Remotec (Ed.).. *Andros Mark V-A1* (Unmanned Vehicle Systems). Retrieved from REMOTEC A Subsidiary of Northrop Grumman: http://www.remotec-andros.com/bodyPages/static/html/androsmva.htm

Space and Naval Warfare Systems Center.. *ROBOTICS.* Retrieved May 11, 2003, from http://www.nosc.mil/robots/

Super70s.com (Vol. Ed.), *The Meltdown of Three Mile Island* [Special issue]. Retrieved from Super70s.com Where the 70s Never Ended: http://www.super70s/News/1979/March/28-Three_Mile_Island.asp

Werner, C.. *Robots at the WTC* (Technology and the Terrorist Attacks: Part 3) [Special issue]. Retrieved , 2001, from Firehouse.com: http://www.firehouse.com/techzone/

About the cover: The robot pictured on the cover is titled <u>Robot talking</u> by *JoannaZh@dreamstime.com.*

ABOUT THE AUTHOR

Jordan Larsen is an avid reader and writer. He was homeschooled from 2nd grade through high school. Since graduating from college in 2005, he now spends most of his time studying the field of emergency management and the role of emergency first responders. He also writes reviews on amazon.com critiquing various soundtracks, movies and novels.

CPSIA information can be obtained
at www.ICGtesting.com
Printed in the USA
LVIC04n0230290815
452039LV00002B/11